SCIENTISTS, INVENTORS, AND TINKERERS

THE DISCOVERIES AND INVENTIONS AS PRECURSORS THAT LED TO PHILO FARNSWORTH'S INVENTION OF TELEVISION

DONALD RAY SCHWARTZ

NVU
NVISION US

SCIENTISTS, INVENTORS, AND TINKERERS

ISBN: 978-1-966782-32-2 (Paperback Edition)
ISBN: 978-1-966782-31-5 (E-book Edition)

Printed in the United States of America

Table of Contents

Preface

Introduction, Scope, Purpose of the Monograph, and Credits

In his Nobel Prize acceptance lecture, Guglielmo Marconi stated, "The discoveries connected with the propagation of electric waves over long distances and the practical applications . . . have been . . . . the results of one another . . . After . . . preliminary experiments with Hertzian waves I became very soon convinced that if these waves . . . could be reliably transmitted and received over considerable distances a new system of communication would become available possessing enormous advantages over flashlights and optical methods . . ."

Marconi was not the first to discover wireless communication. But he was the first to develop it successfully. He did so through trial and error. Thomas Edison favored the trial and error tinkering technique in his panoply of pragmatic inventions that also changed forever the civilized world, including the motion picture camera and projector. Marconi's sea change invention was the very model that sent and received coded and voice communication ship to ship, ship to shore, and, ultimately, the working model for radio.

Throughout the nineteenth century, the elements leading to electromagnetic wave propagation and the mechanical and chemical elements leading to the recording of visual imagery, these seemingly separate fields charged toward one discovery: to transmit and receive voice and picture together. Marconi's success in wireless voice transmission set in or increased the search for lip synchronized talking pictures—an electronic or mechanical media holy grail.

Indeed, in April 1926, a full year before Farnsworth developed true television, John L. Baird announced he had invented the device, which he called the "Televisor."[1]

Even a full century earlier, this quest had become manifest. Again the road to its realization divided into two paths.

Upon one path scientists developed work with colleagues, developed insights in conflict with colleagues, and sought each other's discoveries. They produced the current of electricity, the control of this power, and the manipulation of its forms from which Television emerged. As these scientists persevered through laboratory experimentation and experimental oscillations of the phenomenon, they became most fascinated by the appearance of an eerie luminescence, a beauteous and mysterious glow within their glass tubes. It is a claim by the several supporters of the schools of Plucker, Braun, and Crookes that each champion dubbed this glow's presence Cathode Rays.

Upon the other path inventor-artists developed the concept of using still photographs in rapid succession to acquire the illusion of movement in pictures and of moving pictures that talked. It may not be mere coincidence that these paths converged in 1927.

The scope of the present work is to follow these two pathways to their confluence. This convergence resulted in visual movement and speech in synchronicity. The two advances occurred virtually at the same time. Nonetheless the ultimate synthesis was Philo Farnsworth's invention.

Although considerable amount of material by scholars and writers exists concerning the invention itself, little is available to demonstrate its linear evolution: Of brilliant suppositions; Of ingenious conclusions; Of false starts; Of rethinking; Of flawed attempts; Of success that propelled the charge along—this progression that allowed the inventor to piece the elements together for his fateful "aha" experience one day while he was in his farm field upon his tractor dragging his mower and baler behind.

I propose to follow these two paths, forming the necessary basis for the invention.

As a monograph, the limitation upon the study can only produce an outline of the development. The present writer or a future scholar may wish to develop a more complete investigation.

Karl Braun, who shared the Nobel Prize with Marconi, said in his lecture: "I set myself the task of obtaining stronger effects from the transmitter."

So did they all.

For this treatise and its humble results, I wish to thank the fine librarians at The Community College of Baltimore County, at all campuses, but especially the Essex campus, in particular Michele Meisart, Wendy Sears, and Joan Donati, without whose assistance in obtaining somewhat rare research materials and books through interlibrary loan, I could not even have begun this work. The fine librarians at Towson University deserve mention, especially Shana Gass who also in her institution assisted me in retrieving historical documents.

I would be remiss if I did not thank those officers of the college, who encouraged me to develop this project and determined that it would realize perhaps somewhat of a decent contribution to the field—F. Scott Black, Mark McColloch, and Sandra Kurtinitis. To the members of the Board of Directors who also sanctioned the project, I also offer my humble gratitude.

And to Ann, with whom I watch a few good programs and who takes a very pretty picture.

DRS, Baltimore, Maryland, January 30, 2009

The Charge

Prelude

Although the phenomenon of attraction by a mysterious force had been known for some time, the beginning of disciplined investigation into this natural and seemingly supernatural occurrence can be traced to 1746. Thirty years before the Declaration of Independence Benjamin Franklin began his experiments. With the famous kite experiment and others of his design accomplished, Franklin published two remarkable suppositions. These fundamental concepts have held up unto today as descriptors of electricity: It has a flow, called current; current exhibits positive and negative characteristics.[2]

Even before Volta's pile, Franklin assembled a rudimentary battery. It produced sparks, accommodating a fanciful and thrilling entertainment of the time. It would be some decades before full electric generation.

Even so, Franklin used his shocked-gained healthy fear of the sparks and of the flow, in concert with his famous kite experiment to invent a device that would benefit homeowners and commercial properties then and now—the lightening rod.

In 1780 Luigi Galvani, by stimulating dead frogs' legs and observing them twitch, came to believe that animal tissue (and, thus, that of people) could generate or, at least, conduct, electricity.

In 1799 Alessandro Guiseppe Volta deduced that two metals—steel and tin—upon which the frog rested (thus the tissue connected them) had generated or could generate electric flow (current). This led to Volta's invention of the first true battery, the Voltaic Pile or Cell. Volta's invention, once he placed a conducting wire, was the moment it was determined electricity could be generated and conducted (controlled) in a closed circuit. Andre-Marie Ampere in 1820 and Georg Simon Ohm in 1826 completed the triumvirate begun by Volta, as they explained severally the manner of current as it encounters and acquires—actually, requires—resistance.

Thus, Voltage generates amperes (current) whilst an object desired (a light bulb, a vacuum cleaner, a television set) is dependent upon providing ohms in a completed or closed circuit, the basic principle understood at that time and today as Ohm's Law.[3]

Thus the stage was set for the linear advances that would lead to the world of the twentieth and twenty-first centuries. Throughout, there were many other scientists and inventors than those imposed by the limitation of this study. That is, we concentrate herein on those whose results delineate the inexorable march toward Farnsworth's invention that changed the world—the scanning Cathode Ray Tube[4] and those whose results produced at the same time lip-synchronization talking motion pictures.

Chapter 1

Electric Waves

Michael Faraday

Although it had been suspected by Hans Christian Oerstered as early as 1820, it was Michael Faraday who, a decade later, developed the principle upon which, excepting the battery, all electrical production is based. To develop the magneto, the necessary invention for electricity to be generated on a large scale, Faraday reasoned and placed into practical effect this connection.

"Nowhere is there a pure creation or production of power without a corresponding exhaustion of something to supply it," Faraday said. His impeccable reasoning represents in its simple elegance one of the most important scientific discoveries: If electricity can produce magnetism, magnetism can produce electricity.

From this reasoning, it was a short series of steps until the wound coil, a generator, was produced. From here, the reasoning continued—if electricity can generate magnetism, and magnetism, it was known, produced lines of force, therefore electricity produces lines of force.

In the latter part of the nineteenth century, Faraday's "lines of force" would become known as electromagnetic wave radiation, the controlled process for the production of radio and television (and, in our own day, the telephone).

But Faraday's contribution, as immensely significant as it is, does not end there. Years later, in 1840, Faraday discovered the link between light and electromagnetism. The Faraday Effect, as it is called, is complex; however, in essence, a beam of light is shone through an opaque light sensitive apparatus of a specific conductor design, and stimulates a magnetic field.

Without discussing other implications, Faraday, prior to light appearing within glass tubes, demonstrated that light can be generated and can generate an electromagnetic flow. The search for light and its imagery would be forever linked with electromagnetic generation, again, later known as electromagnetic wave radiation.

Faraday also coined the terms that the scientists, inventors, and tinkerers would need to develop their own work and design their own experiments: Electrode; Anode; Cathode; Ion.

Michael Faraday was born on the outskirts of London in 1791. Due to his father's chronic illness, the household was impoverished. At fourteen, the boy was apprenticed to a bookbinder. Books and treatises entered the shop, enabling the aspiring scientist to sate his appetite of reading voraciously the subjects of science and electricity which consumed his interest.

As a young man Faraday attended the renowned lectures of Sir Humphrey Davy at the Royal Institution. Within a short time, Davy, impressed with Faraday's extensive self-taught scientific knowledge and his interest in this new discovery ensured he had an apartment and a laboratory.

Throughout his life, in this domicile and work space Faraday conducted his history changing experiments.

Faraday retired from laboratory research in 1855. By demand he would be brought back to lecture until 1861. He died in 1867 at the age of 76. He is buried in Highgate Cemetery in London.[5]

Johann Heinrich Wilhelm Geissler and Julius Plucker

Johann Heinrich Wilhelm Geissler received a tradesman's apprenticeship, a journeyman's award, and the accomplishment of a master; but he acquired no scientist's education significant to discern the nature of electricity. Of course, he could not know that the elegant materials he produced ultimately would lead to television. But he had the serendipitous good fortune to be acquainted with a physicist most interested in Faraday's work, Julius Plucker. It is this chance and developing relationship that led to the precursor of the very picture tube itself.

Geissler's family business had been glass blowing. How far back the family's trade extended, even the members of the family didn't know for certain. Glassblowing is one of those crafts that is very old. It extends back even into the ancient near east. It has been largely practiced in much the same fashion through the ages, even unto today. So Geissler may have gone on, producing his familiar wares.

But in 1852, Geissler was approached by one of his countryman—a mathematician and physicist. The scientist stated an odd request.

Still, Geissler had known many strange orders. His reputation for his skill was probably what prompted the scientist to knock on his door.

Julius Plucker had already established his reputation as a leading mathematician and scientist of the day. In 1847, Plucker received an appointment of Professor of Physics at Bonn. About this time he became interested in Faraday's work. Faraday had begun to experiment with electrical discharge in gases, noting the spark effect. For some reason, Faraday did not pursue a manner to control the spark effusion. It occurred to Plucker that if the gases could be contained in an enclosure, the discharge effect should be observable for a length of time.

At first Geissler blew to create small devices. In time, they received practical expression as thermometers. Finally, in 1858, Geissler invented the vacuum glass tube. When Plucker generated the discharge, an eerie, mysterious, and beautiful greenish glow appeared. It remained for a persistent time.

Geissler had invented the eponymous Geissler Tube. It would evolve into the Cathode Ray Tube. It would become the picture tube of television. Plucker the complete scientist recognized that this fluorescence within the tube responded to an electromagnet on the wall of the tube; he discerned that these expressions of light were rays or beams of some electrical property.

Crookes would later refine and identify the Cathode Ray Tube; Thomson would understand these rays were electron beams. At this time, however, Geissler and Plucker had propelled Faraday's Effect into a visual manifestation. For that, the game was afoot.

Heinrich Geissler was born in Igelshieb in Saxe-Meiningen, Germany in 1814. He died in 1879.

Julius Plucker was born in Elberfield in 1801. Prior to his appointment at the University of Bonn, he held lectureships at the University of Warburg, University of Halle. Plucker died in 1868. The next year, a student, Johann Hittorf, continued his professor's work and demonstrated that the phosphorescent glow must be rays, coalescing as they were formed from one end, propelled through the length of the tube.[6]

James Clark Maxwell

One of the greatest minds of the nineteenth century never worked in a laboratory to develop his concepts. Yet this mere theorist presumed to investigate Faraday's electromagnetic force, and the

3

Plucker school's identification of Cathode Rays literally to shed light on the entire phenomenon. His resulting proposals transformed even modern and contemporary thought.

James Clark Maxwell emerged as the singular brilliant mathematician of the nineteenth century. Between 1850 and 1856, he immersed his mathematical investigations into this new and challenging arena of electromagnetic wave propagation. Through a series of highly complex and sophisticated mathematical formulae, Maxwell developed a startling conclusion. He realized the possibility of a unified field theory.

Using only his powers of induction and deduction, through higher mathematics, deriving himself some of the theorems, Maxwell calculated his ultimate series of equations. They described nothing less than two theories which gave the scientists, inventors, and tinkerers who followed him the bases for their own work: Electromagnetic waves propagate through space; Light consists of electromagnetic waves.

They proved Faraday's suppositions. Through his publications they established a proof-text. They provided a theoretical clarity to what occurred visibly within Geissler Tubes.

Vast implications spread at the conclusion of the formulae. At last, scientists and experimenters realized potential answers to their questions concerning the curious nature of the effect, when the voltage was raised. At last this profound realization was understood: Light consisted of this same process; all electromagnetic waves race at the speed of light. Thus Maxwell achieved, through higher mathematics alone, a dynamic theory of the electromagnetic field (1864). Though the equations are quite complex, understood fully only by those trained and read in higher mathematics, Maxwell elegantly expressed their treatise coherently: "We can scarcely avoid the conclusion that light consists in the transverse undulations of the same medium which is the cause of electric and magnetic phenomena."

Years later, Albert Einstein considered Maxwell the genius non pareil of the nineteenth century, from who all else comes; he stated that without Maxwell, he could not have developed his Special Theory of Relativity. The scientists and inventors who followed Maxwell placed so many heritages in his work they referred to themselves as Maxwellians.

James Clark Maxwell was born in Edinburgh, Scotland, in 1831. A prodigy, he attended the Edinburg Academy when only eleven years of age. In 1850 he attained an appointment at the University of Cambridge. Later he lectured at Trinity College. Maxwell was appointed to King's College in 1860. Maxwell died in Cambridge in 1879, never seeing his theories, equations, and conclusions put into practical use.

But those who followed him knew to whom they owed their own discoveries.[7]

Heinrich Hertz

"I have attempted to demonstrate the truth of Maxwell's equations . . . (I) infer without error that . . . (they are) certainly to be preferred."

Heinrich Hertz wrote this humble yet startling conclusion in 1884, having realized the nature and profound meaning of his discovery. For Hertz was the scientist who produced what Maxwell had predicted— electromagnetic waves, waves along the entire spectrum of invisible lines of flux and force. These were the waves that could produce wireless communication.

Through his experiments, Hertz demonstrated wave properties as wavelength, reflection, interference, refraction. With this spectrum revealed, within a few decades, those who followed produced amplitude and frequency modulation for wireless communication, AM and FM radio, and, ultimately, television.

In 1879, Hertz won a prize sponsored by the Berlin Academy of Science's Philosophical Faculty. The sponsorship was interested in determining whether electric current possesses mass. Hertz did not reduce his findings to an answer of this question (none of the contestants were able to); but the manner in which he conducted his investigation determined his victory.

What he had done was to modify a contemporary apparatus to produce effects not seen before. He published a paper from this series of findings: "Kinetic Energy of Electricity in Motion."

By 1883 Hertz moved into a new laboratory. In the interim Hertz had studied Maxwell's equations and conclusions. He set himself the task to realize in practicality the master's theorems. He increasingly became amazed and comforted perceiving he seemed to be the only scientist in the world engaged in this pursuit. He was!

Interestingly, he was teaching a class of students when the moment occurred. It was what would become standard in any class of college physics, continuing in classrooms today: A demonstration designed to amaze and inspire students—spiral coils that produced sparks across their gaps.

Certainly Hertz knew well of Faraday's findings as well as Maxwell's equations.

However, through a series of circumstances not quite known, Hertz had to place his coils farther apart than he desired or had prepared in his pedagogy. Thinking it would never work, he was surprised when the second coil, farther apart than he had designed in his lesson plan or had wished, reacted with an induced current. That is, the spark and current flow in the primary coil had induced current in the second a considerable distance apart.

Hertz at once recognized a new line of investigation opened, as a previously unopened book, before him. In the main, how far apart could the induction coils be when current would still be induced successfully and undeniably from the primary to the secondary?

For awhile his puzzlement over the small voltage kept him from his ultimate conclusion. When he realized that no matter how he manipulated the coils and the spark gaps, current was induced, he finally came to the conclusion that only waves traveling across the distance could be the cause. By 1887 he was certain.

He had uncovered the propagation of electromagnetic waves. Now, through his more sophisticated designed experiments, he understood the waves could traverse walls. Perhaps they might travel a distance, perchance conquering even an extended geography. They may cavort over hill and dale. He also realized that a manifestation of light propagated and was fully related to these waves—just as Maxwell had elegantly predicted. Indeed, Hertz went on with this realization to discover the photoelectric effect, by concluding that ultraviolet light made the spark gap more conductive on a device he called quite logically the detector.

Hertz also became most interested in the work of Eugen Goldstein, who was developing the Cathode Ray Tube.[8] Perhaps but for his early death, Hertz, with his accumulating insights from his investigations, might have brought about the radio or television age sooner. But perhaps not—for Hertz, as this study reveals of scientists again and again, was interested merely in the process of discovery. That is, his intense curiosity alone drove him to experimentation and supposition. Interest in pragmatic application beyond theory did not often occur to him.

Heinrich Hertz was born in Hamburg in 1857. At an early age he already excelled in science and studied the discipline along with Arabic and Sanskrit at the University of Hamburg. He studied under Hermann von Helmholtz, who counted among his other proteges Albert Michelson, who first measured the speed of light, and Max Plank, who began the discovery of quantum physics. It was a talented department, to be sure. Hertz was encouraged in his pursuit by his esteemed professor, truly a discoverer in his own right—of talent at the highest level.

Tragically, Heinrich Hertz died from blood poisoning brought on by a jaw infection in 1894. He was thirty-seven years old.[9]

Oliver Lodge and William Crookes

Oliver Lodge's biographers and Lodge himself credit the scientist the first to put into some pragmatic effect Hertz's theories and suppositions. Only a few months following Hertz's early departure from this mortal coil Lodge prepared his reverberating demonstration.

Even before Hertz identified the phenomenon, it had been used in America as early as 1842 and in the 1880's to an admittedly lesser extent by Granville Woods, Thomas Edison, and Lucius Phelps, enabling stations briefly to communicate with close passing trains. Albeit workable, these were feeble attempts. They were not always dependable. Soon these more primitive inventions were discarded.

With Hertz's full identification of the principles and potentials inherent in the phenomenon, Lodge set about to design for public view the practical effect that Marconi would read about, so becoming stimulated to begin his own inventive process.

Lodge had invented a device he called a coherer. Simply put, metal filings would cohere in the secondary coil when the primary coil was activated across the room. The flashing spark that began the experiment resulted in full success chillingly accompanied by a resounding boom. Marconi later relied upon this spark-induced thunderous burst for his initial endeavors. Hertz's theories received instant validation. After all, they were visually apparent. Lodge should continue his work. There was speculation he might produce wireless voice transmission, perhaps even visual transmission.

Alas, Lodge harbored a certain tendency, or, ironically, perhaps, an uncertain tendency. Once he proved or came close to proving a principle, he turned his attention to a different focus of wireless communication—searching for a way to transmit and receive messages with the departed. Or, if not involved in investigations of the arcane, Lodge would precede typically to other scientific investigations—that is, the next project. Although Marconi worried that Lodge would beat him to the punch, he need not have been so concerned.

Oliver Lodge was born in 1851 at Penkhull, Stoke on Trent. Lodge was a sibling of a large family but his father was quite successful at business, so the many mouths were fed with little or no concern.

As a young man, Lodge enrolled in King's College in Mathematics, Mechanics, and Physics. He soon became most interested in electricity, by now of itself a well known phenomenon and useful principle.

In 1881 he received appointment as Professor of Physics at University College, Liverpool. Here was his laboratory that produced considerable influential work for twenty years.

Oliver Lodge died in 1940, a long and fruitful life lived on this sphere and searching for those beyond. His several investigations plowed groundwork for future discoveries, including Einstein's. They exceed the limitations of this study. For this purpose, however, his work formed the basis for the later development of long range wireless telegraphy and radio, his coherer eventually evolving into the vacuum and audion tubes.[10]

If Lodge may have come close to inventing wireless and radio, Sir William Crookes, Lodge's contemporary colleague and friend, may have just missed discovering television.

Crookes's interest in the Geissler Tube and Cathode Rays occurred early in his scientific career. Although his investigations of the apparatus and his attempts to uncover the mystery of its phosphorescence did not occupy solely his attention, he continued to return to it. About the same time his colleague and friend was proving Hertzian Waves, he identified and even invented the device that would be instantly recognized today—the (precursor of) the Cathode Ray Tube.

As early as 1878, Crookes wrote in a letter to his early collaborator, George Gabriel Stokes: "I have been constantly thinking about the ultra- gaseous ... state of matter in the "green" vacuum ... (Perhaps) the . . . space is filled with ... molecular matter ... a chain of molecular impacts may carry the force along in a straight line ... the ... colour ... is the phosphorescence."

Though even he did not fully recognize this implication until later, Crookes had determined that it is a stream of energy from one point of the tube to the other that illuminated the glass. Later, when Thompson discovered or was the first to induce the reality of the electron, Crookes understood this stream well. He referred to the stream as them, as though predicting their revelation, electrically charged particles.

Although Plucker and Hittorf and, more so, Goldstein, had recognized and coined the term Cathode Rays, what Crookes had done is to identify definitively what occurred in the tube. It could now be designed to produce the effect in a more controlled, perhaps even desired fashion.

Further, Crookes analyzed the process within the tube so fully, he was able to map it. That is, he developed tubes which operated more effectively. Almost as if "tuning" them, he discovered he could, to an appreciable extent, control the phenomenon. The handle these objects went by then and go by now is the eponymous nomenclature of "Crookes Tubes."

At one end of the tube, say looking at its left side, was the Cathode. The voltage then cast the particles through a mysterious, unlit space, again receiving an eponymous name, "Crookes Dark Space"; thence moving right, not yet at the center of the tube, a hesitant but constant illumination, termed the "negative glow"; more toward the center is observed the curious darkening Faraday once noted, also appropriately identified as the "Faraday Dark Space"; from here, then, proceeding to the Anode. The illumination took on what appeared to be striations moving one after the other. This appearance was termed the Positive Column.

Today the precise manner of the necessity of these several parts and how, at the sub-atomic level, the energy distribution for the production of electromagnetic wave radiation in the spectrum of light is realized is more fully understood.

Crookes now had everything he needed to develop the invention further. Like his colleague Lodge, however, he set off in pursuit of communicating with the dead, in seances and other supernatural venues.

Yet his many achievements and this one in particular, established precisely what was needed for the telegraphy of instant pictures.

In 1892, two years before Lodge's world changing demonstration, Crookes observed:

... electrical vibrations of a yard or more in wave-length will pierce (walls). Here is revealed the bewildering possibility of telegraphy without wires, posts, cables ... an experimentalist at a distance can receive some, if not all, of these rays on a properly constituted instrument...

Did Crookes's analogous interest in wireless inhibit his further investigation and development of the Cathode Ray Tube? It is an interesting question, which must remain part of the enigma of William Crookes.

In any event, as Lodge and Crookes diverted their fancy to their more arcane investigations, the field was then open for the man who would at last cross the boundary and produce wireless and radio.

Sir William Crookes was born in London in 1832. In 1854 he received appointment at the College of Science in Chester, Cheshire. Soon thereafter he came into his father's considerable estate. With this fortune, he was able to develop his laboratory. Through the late 1880's into the last decade of the century he used this lab to conduct his Cathode Ray Tube experiments and investigations and in his attempt to penetrate the world beyond. Perchance he saw them as one and the same.

Crookes died in 1919, nearly a decade before his tube—rather the inventive and incisive modification of it—produced the first true televised image.[11]

Guglielmo Marconi

In 1844, a message was sent instantaneously from Washington to Baltimore. Anne Ellesworth requested the transmitting operator tap out the series of dots and dashes that produced the words, "What hath God Wrought!" Samuel F. Morse had invented the telegraph. Within a few years the familiar poles and wires criss-crossed the country and soon the civilized world. The profound nature of this achievement cannot be overestimated. In 1776 John and Abigail Adams's correspondence, only a few hundred miles apart in Boston and Philadelphia, could take days and in winter, even weeks.

Communication across the Atlantic might take months. More often than not, letters crossed. Before the middle of the next century, messages could be delivered within hours, even, for all practical purposes, instantaneously. Morse's invention changed the world.

From the middle of the century into the time of Lodge, Crookes, and Thomson, especially since Hertz's discovery, the search progressed for this instantaneous communication, without the inconvenience of cords or wires. As we have seen none of the scientists, brilliant and ingenious and important as their discoveries were, developed the practical model, nor seemed concerned to.[12]

Guglielmo Marconi admitted he was not a trained scientist. Like Edison, he was an inveterate tinkerer. As Edison invented the light bulb in a series of stops and starts, so Marconi invented the first working model of wireless over considerable distances through trial and error.

Marconi followed the scientific news of Lodge's spark and electromagnetic wave radiation from a transmitter to a coherer (receiver). At once he recognized that if wireless waves could be sent across a room, a larger spark and larger equipment could transmit wireless messages across the countryside, perhaps across the ocean.

Toward the end of the same year of Lodge's spark-gap transmission, Marconi built his own apparatus. He may have been aware that in 1892 William Preece almost by accident had discovered induction wireless communication. Preece was able to transmit a little over three miles. Later Preece would welcome Marconi to England, recognizing the Italian was on the right track.

Although Marconi at first reasoned that the larger the spark—the further the waves could propagate, thus setting "thunder," as it came to be called, he induced that the larger his transmitter and receiving wires, the larger the range. He soon found this was so. Incredulous that no one had thought of it before him, he had invented the two necessary components to transmission and reception—the transmitter tower and the receiving antenna. In 1895, he transmitted a wireless message—the Morse Code "S." The dots and dashes transmitted several miles, including over hill and dale. The transmission was received. Marconi knew it was time to patent his invention. Recognizing the greatest potential probably lay in maritime communication, Marconi departed for England in 1896. The Italian ambassador introduced him to Preece, now Engineer-in-Chief of the General Post Office.

Upon witnessing the effect Preece became convinced. Marconi received investors' funds (they would realize excellent returns); he began with naval and commercial shipping's intense interest in his expansion and perfection of his invention.

In time, again by tinkering he recognized how to design the coil of magnetized wire, and perhaps his ultimate discovery—that the increasing size of the towers and antenna didn't matter as much as their architecture —the complex "map" of the wire design. The spark effusion became less important, the coils and intricate antennae design more so.

In 1926, as television was being invented, Marconi, in a special to The New York Times, recounted 25 years earlier his transatlantic success. In other words, in only 7 years, he had gone from his bedroom experiments in Italy to ship to shore and ship to ship communication to transmission and reception across the ocean. Even the transatlantic cables resting on the bottom of the ocean floor to convey telegraphs had become then obsolete.

"In 1895 and 1896 I had proved the possibility of transmitting signals to a considerable distance by means of raised antennae and an earth connection. In 1899 I had proved that the curvature of the earth did not interfere with a propagation of ether waves over short distances and in 1909 I felt that the time had come to venture further afield. Having regard to the many improvements I had lately introduced into the methods of tuning the transmitter and receiver, I was absolutely convinced that transatlantic wireless telegraphy, not merely as an experiment but as a sound commercial proposition, was possible."

On Dec. 12, 1901, in a room of a disused barracks on Signal Hill, St. John's Newfoundland ... on a table stood some instruments .. connected (to) a telephone, in which shortly after noon were heard sounds constituting evidence that in far-distant Cornwall rhythmical signals that corresponded to the letter S in the Morse code had been projected into the ether of space and had actually crossed the Atlantic.

Over the course of that 25 year period, wireless communication evolved into radio and the beginning of television. But it was Marconi the inveterate tinkerer who had brought wireless connection to the world, and began the process.

Guglielmo Marconi was born at Villa Griffone in 1874. However much of his life and career was spent in English speaking countries, England and America. His invention was always scrutinized by military general officers.

Toward the end of his life, he returned to his homeland. With his wealth (for, unlike the scientists who progressed mainly for intellectual curiosity, Marconi, like Edison, ensured his patents and his investors' returns) he lived in a sumptuous Rome apartment complex. Unable or unwilling to perceive fascism as evil, probably due to his resurgent ethnocentric patriotism, he aligned with Mussolini. However, he died in 1937, a full two years before war broke out, and the road to catastrophe began. He had witnessed the full development of radio and the fledging origins of television. His funerals (there were two) were, as Lodge and Crookes might have opined, (Karl) Ferdinand Braun "I found in 1902 that an antenna, inclined at somewhat less than 10° to the horizon, formed a kind of directional receiver . . . a clearly defined maximum for waves ..."

"... the Cathode Ray Tube which I described in 1897 ... provided a visual picture of current—and voltage-waveforms . . ."

So Karl Ferdinand Braun described in his Nobel acceptance address his two contributions that led to radio and television. Braun received his prize the same year as Marconi. It was a shared prize. The physicist has walked in the shadow of the tinkerer ever since. Perhaps Lodge and Crookes in the afterlife are aware of his frustration. In any case, two things are evident. Marconi wears the rightful crown as the inventor of wireless and radio; radio and television as we know them could not have been invented or developed without Braun's discoveries. In his presentation Braun recounted his improvement of Marconi's system. He invoked his invention of the oscilloscope, still used by electronics technicians today to diagnose equipment malfunction.

Braun was appointed Professor Extraordinaire of Theoretical Physics at the University of Marburg in 1876. In 1885 he established or helped to establish the new Physical Institute. Here and at other labs Braun became fascinated with the Geissler apparatus and Crookes Tubes.

As he wrote, by 1897 he recognized that the CRT could phosphoresce a direct image. Braun modified the shape of the tube into the narrow to wide profile, one that has become an icon of the modern and contemporary world. The energy produced propelled the electron wave from the narrow end toward the larger glass at the opposite end. As the device fired this controlled beam of energy coalesced as line images of voltage and current (amperage), appearing in a manner precisely reflecting the charge upon the glass. At last, a controlled image, an image oscillated with facility danced in interpretive fashion upon what would soon be labeled the screen.

13

Although no one would think of scanning the electron field left-right, right-left in the horizontal plane for over twenty years, the advance to allow the full step to the invention of television was now complete.

Braun also, as he stated in his Nobel address, came to recognize the architectural nature of transmitting and receiving antennae. Again, although Marconi always claimed, and rightfully so by the evidence, that he had improved his own system in a same or similar fashion, with little doubt, Braun's improvement of the antenna design was a leap forward as the age of radio began.

In one of the few instances of a rare scientist perceiving the economic practical value, Braun emerged as Marconi's technological and economic rival for the early decades of the new century. Braun was among the founders of the Telefunken Company. He engaged Marconi in numerous patent disputes.

Karl Ferdinand Braun was born in Fulda, Hes se-Kessel in 1850. Even as student, his insights were recognized early at the Universities of Marburg and Berlin.

After the outbreak of World War I, Braun was called to the United States to testify in the matter of one of the several patent suits. The court case was decided as the war erupted in full fury. Braun could not return home. He died in the United States in 1918.[14]

Lee De Forest

The end of the nineteenth century and the beginning of the twentieth brought changes and inventions unimaginable even as late as 1875. The automobile, the washing machine, the airplane, air conditioning, the calculator that led to the computer—all in the times that were changing. Suddenly there was radio; then television transfixed the populace.

Suddenly radio, following Marconi's landmark invention, is the story of three inventors: Reginald Aubrey Fessenden; Sir John Ambrose Fleming; and Lee De Forest. Including Marconi they conducted their experiments, investigations, and insights at or about the same time; at once each improved upon the other's advances.

There are testimonies that Marconi heard voice more than once, albeit faint. For Fessenden's transmissions more reliable evidence can be found.

Fleming invented the vacuum tube. Without this tube, or series of tubes, radio and, later, television could not have developed. Fleming's valve—or the vacuum diode, converted alternating current wave signals that could be heard with a telephone receiver.

In 1906 Fessenden used a high speed alternating current generator to transmit speech eleven miles from his remote location to a receiver at Plymouth, Massachusetts. The age of radio was born.

Still, there was limited use beyond what Marconi had already developed; after all, the inventor of wireless transmitted and received with regularity messages across the Atlantic Ocean. The fledgling radio transmissions, not at first clear, could not process with distinction the distance from Baltimore to Washington, a litmus test since, it will be recalled, wherewith the first undeniable telegraph transmission had transpired.

Lee De Forest experimented with radio as early as 1903. A graduate of Yale, he had always been interested in wireless concepts. He studied Edison's work and Marconi's invention intently. Through jobs at AT&T and other companies it occurred to him that the diode vacuum tube, if increased to a triode, could amplify and perhaps even clarify the sound of speech and music. In the same year Fessenden transmitted, De Forest added his third element to the tube. At first the results were not spectacular. Within a few years, however, it was clear that a new medium could transmit through air voice and music across the country. This new arrival would change America and world life forever. It was one thing for long range telephone communications between two persons, or even Marconi's remarkable invention between one station and another. What De Forrest did was promulgate a new mass communication module. The voices live and recorded presentations now went at once to hundreds, thousands, millions—even more than the daily newspaper—and that, simultaneously.

In 1920, Station KDKA in a garage in Wilkinsburg, Pennsylvania, began producing and transmitting. By 1926—about a year before the invention of television, David Sarnoff—who had begun his career under Marconi's tutelage—formed NBC. Sophisticated studios broadcast to thousands of radio receivers.

These were of De Forest's lineage—a series of triode vacuum tubes accommodating amplitude modification. There was static to be sure, but news, weather, sports, music and theatrical dramas magically coursed through the air to businesses, families, politicians, and lonely hearts.

Unfortunately, as brilliant an inventor as De Forest was, unlike Marconi and Sarnoff, he lacked wisdom or acumen in business. He called his invention the Audion Tube. He sold it and the subsequent rights to AT&T for $50,000. It is estimated the company and, in turn, RCA/NBC realized untold billions.

Lee De Forest was born in Council Bluffs, Iowa in 1873. His life of brilliance and difficulty began in childhood. The family soon moved to Alabama, where his father was appointed president of a black college. De Forest never felt at home in either the white or black communities.

Nonetheless, he was a brilliant student. Accepted to Yale, he graduated with high marks and a lasting memory of perhaps his most halcyon days.

De Forest lived a long if uneven life. He died in 1961. He witnessed his invention improve and spread throughout all of American and world culture. Along with his invention allowing lip-synchronization in talking motion pictures, he recognized with assured sense of accomplishment he was responsible for two of the three most influential mass communication devices of the modern and contemporary world.[15]

Chapter 2

The Visual Image

As the evolution of wireless communication and radio, borne from scientific curiosity, reached its several climaxes, it began to occur to the devices' inventors, developers, and successors that perchance pictures could be transmitted in this fashion. This search may very well have been stimulated by the dramatic success of another process that was also gaining in momentum and practicality: The production of moving pictures, at first in eyepiece peephole apparati—later projected upon a screen in theaters.

The search for voice and moving pictures caused to appear by a few upon a surface to the visual appreciation of many may be nearly as old as our species walking the earth. From 10,000 to as far back as 40,000 years, men and women painted brilliant images on their cave walls. The most famous of these, dating to about 25,000 years ago, transpire in the various cave galleries at Lascaux.

As the archaeologists and spelunkers granted entrance have recently discovered, if there is torch light, similar to that lit by our distant ancestors, and no modern illumination interfering, the images appear to move.

... the light travels along the vertical walls and suddenly reveals . . . the bison, its entrails hanging out, its tail lashing the air in rage. A long spear cuts across it. The bison has two manes (,) a device used by the artist to signify the movement of the animal as it lowers its head to gore its enemy. . . . (There appears a) man . . . dressed up in a bird mask, thrashing about with arms that are almost as spindly as his legs. (p. 150)

It became clear to us that the cave art at Lascaux was not made to be viewed like a series of fixed paintings, but as images in movement glimpsed by the Paleolithic initiates as they walked through the dark (with their firelight). The power of the images as they loom out of the blackness as if in a dream, taking one by surprise, would have shown them—as they did us —the revelatory power of the religious and magical forces which fuelled the Paleolithic imagination . . . (183)[16]

The first somewhat true photograph, referred to as a heliograph, since the chemicals involved transformed onto a plate through exposure by the sun—a cityscape—was derived by Joseph Niepce in or about 1816. Louis Jacque Daguerre achieved the full process. A lens aperture opens a certain diameter at a certain speed to allow light to disturb a plane treated with chemicals (silver nitrate was ultimately found to be superior). That plane is thence treated with other particulate chemicals. As if prompted by a magician an image emerges before the eyes.

Today, in our digital age, this process may appear to some passé or may come to be considered a dinosaur. But there are those who claim its pictures yet resonate more fully and richly. Daguerre's world changing process came into focus in 1839.

Almost at once, even as still photography advanced the search for pictures that moved was joined. Louis Jacque Daguerre was born in Corneilla¬en-Parises.[17]

Sontag has stated nothing captures the moment in time as does the still photograph. Nonetheless, pictures that moved became the object of desire of many—but three inventors and tinkerers realized the success of the mechanism and the aesthetic—four men, actually, as one unit consisted of two brothers. In Britain, France, and the United States, independent of each other, the same theoretical concept developed that would create moving pictures to be viewed on a projected screen—our dream vision, as our far distant ancestors had perceived theirs.[18]

The three nations lay claim severally that their man (men) invented the motion picture camera and projector. As three filaments energized one glowing and transmitting audio tube, so all three claims may be right.[19]

That the seekers of immediate voice and moving images transmitted by wireless to receivers everywhere were influenced in their search by the magic of moving pictures and sound motion pictures should not present a hyperbolic argument.

There may, after all, be ether: Great ideas tend to flow simultaneously within it.

William Friese-Greene, the Brothers Lumiere, and Thomas Alva Edison

Like most time changing inventions, what seems a complex problem, once the solution develops, appears simpler. The concept of film (pictures that move) relies upon a Gestalt phenomenon, clearly expressed as Persistence of Vision. That is, if a series of still pictures is shown to us at a certain speed, the rods and cones upon our retinas maintain for a split second the stimulus of the image as the new picture proceeds into view, thus giving the illusion of motion. Thereby, film itself is the ultimate special effect.

The effect can occur in as slow a rate as 12 pictures or frames per second (fps). The ideal came to be discerned as 16-18 fps for silent film, 24 fps for sound film. Today, digital cameras and projectors giving the appearance of film are increasingly used (granting an illusion of an illusion, as it were). Still, these devices, perhaps like television, would never have occurred unless Friese-Greene, the Lumieres, and Edison had not developed the original mechanism, predicated on this ophthalmologic aspect of our neurophysiology. In a sense, we are each a television transmitter and receiver. We rely upon the electro-chemical nature of our optic nerves and occipital lobes to respond to the sound and visual stimuli of the world.

The mechanism, then, is that a looped reel of film is pulled singularly by its sprocket holes from the home reel to the receiving reel whilst proceeding through a gate behind which is a powerful light source, in effect, a magic lantern projection.

William Friese-Greene may have been the first to realize this concept. A photographer's assistant, he became obsessed with the idea that a series of stills filmed, then played back in that device could produce the illusion. He was particularly interested in Muybridge capturing horses galloping in a series of photographs, clearly suggesting movement.

Indeed, as early as 1873, Friese-Greene had solved the problem of the series of still pictures in sequence. It astonished those who observed it.

In 1887, he used paper images with sprocket holes to project a moving image in a store window. Yet he knew these were only early attempts. The actual product eluded him—until 1889. He had figured it out at last. Celluloid film appeared. Now he could design his camera. The first true motion picture may have been filmed and later projected a scene of couples walking in Hyde Park. Later, as he cranked his projector device, he realized in awe he had brought moving pictures to the world.

American texts and film instructors will state that Edison invented the camera and projector, French texts the Lumieres. Again all are accurate. But the man overlooked throughout the twentieth century as film fully developed as we know it may well have been the first out of the magic box.

William Friese-Greene was born in Bristol in 1855. He died in London in 1921, already viewing his claim and renowned fading.

August and Louis Lumiere presented their invention in 1895, a device which, in essence, established at once the workings and the design of the camera and projector. This original camera, called the Cinematographe, still takes clear moving pictures today.

The advantage of the Lumeire's camera was its relative lightness and the clarity of its lens. The Lumieres are renowned for one other development. They standardized motion picture film at 35 millimeter in its horizontal to vertical format.

Scholars credit the Lumiere Brothers with the first showing of the first film; again, Friese-Greene may have been first. The Lumieres had an advantage—a well heeled sponsor who gave up his "space" to allow the "shoot" on location. *Lunch Hour at the Lumiere Factory* was shown before the Societe d'Encouragement de L'Industrie Nationale on February 13, 1895. By December 28, the Lumiere films were being shown at the Grand Cafe; heralding that particular circumstance which would become all-important in filmmaking, admission was charged, establishing the first motion picture box office and the counting of the receipts. Within a few weeks, as successful filmmakers have concerned themselves since, the Lumeire's and the cafe owner were ecstatic over the flocking public and their rate per screen showing. In addition to their previous oeuvre, they showed at least five other films including the most well known today, *Arrival of a Train at the Station.*

The titles of these films indicate how they were basically home movies as we know them. It is important to note that it never occurred to Friese-Greene or the Lumieres to use their inventions to tell a story. Soon, film recordings of plays would be shown. But film as its unique story telling exponent—later realized in television—would be the aesthetic brainchild of Edison's studio. Friese-Greene may have begun the concept (there is evidence he wrote to Edison describing his invention); the Lumeires are unquestionably the first to project film for a large audience; but the especial art form of the motion picture without question emerged from Edison's studio.

August Lumiere was born in Lyon in 1862. Louis followed in 1864. Their father, Antoine, had long been interested in the new discipline of photography. From his portrait studio father and sons developed the second largest photographic company in the world, improving photographic plates. The business had the background and resources the brothers needed to invent their camera-projector.

August especially lived a long life, witnessing the explosive growth of film and cinematographic technique, as well as the smaller screens in homes. He died in 1954. He followed Louis who had been gone since 1948.[20]

Thomas Alva Edison is known as the premiere inventor and innovator. Without doubt that is true, to a large extent. However, it is important to note that he was the premiere tinkerer and supervisor of good men and true —and improver of their work. He did not invent the light bulb but, through trial and error, perfected it where it would illuminate for far more than a few minutes. He perceived how electricity could be generated on a large scale and distributed; nonetheless he was forever hampered by his obsession with his cherished Direct Current. In actuality it would be Nikola Tesla who, developing Alternating Current, would give us more fully the modern world.

Edison invented recording and playback wound cylinder disks, which led to the phonograph. At virtually the same time as Friese-Greene and the Lumieres, his shop invented the motion picture camera and projector. Probably for patent reasons, he designated the device the Kinectoscope.

It was Edison who helped to develop celluloid film, a problem that George Eastman, in consultation with Edison, solved and began producing in 1895—thus allowing the projected image to travel through the camera and projector gate from its home reel to its take-up reel.

At first Edison produced and promoted his film strips in arcades. A patron deposited a nickel in the Nickelodeon coin slot. S/he peered into the blinder surrounded peephole. S/he cranked the side handle to move the strip from home to take-up reel; or, an attendant cranked at the proper speed. These filmstrips ran from 10 or 15 seconds to a half a minute or more. The viewing public thrilled, amazed, and enthralled, was shocked at the most famous of these strips, which portrayed the first film kiss. It was entitled, appropriately, The Kiss, with the gentleman smoothing out his moustache and beard before pressing his lips to the more delicate pair of the lady's. Within a few years, Edison and others projected their films on screen.

Since the early shutters in the gates lacked the calibration that was later attained, these early images "flicked," hence the terms, "flickers," or "a flick."

Edison had always done well in his shop by hiring loyal, skilled, and talented men, who were innovative, imaginative, and competent in their own right. For many of Edison's over 1,000 patents, it would be more accurate to refer to them as belonging to Edison's Menlo Park operation.

He always knew how to hire the best people for the job. As the film industry grew it occurred to him to engage people from the theater. D. W. Griffith and Edwin S. Porter, in Edison's especially constructed "Black Maria" studio in New Jersey would develop the art of the motion picture, using Edison's technological accomplishment.

Griffith's *Escape From Eagle's Nest* runs about 7 minutes. To today's sophisticated audiences it appears middle-schoolish and bloated with naivete. Audiences then, however, had never before seen film tell a story. Though little evidence exists, Edison and Griffith received reports that women swooned when viewing the handsome film actor larger than life in his woodsman's shirt unbuttoned with planned carelessness.

In 1903, Porter took the camera on location. He shot a 12 to 14 minute film that established the art of the motion picture. For the first time, there was a script with day or night indicated and with instructional elements, directional continuity, cross-cutting, post production editing, a close-up. *The Great Train Robbery* arrived to begin true theatrical film.

"Those anticipating viewing this early work expecting to see performers as handsome and beautiful as those kissing on screen today will be sorely disappointed.

It also established the heritage of the western movie genre. Today it also seems naïve and quaint. But most of the elements were up there, to be seen. Contemporary audiences were astonished.

Griffith went west to be among those who established Hollywood. He and those who quickly followed began the process of producing the more sophisticated cinematographic technique we enjoy and expect today.

Edison may not have been the first precisely to invent the camera and projector that changed the world. But it was he who encouraged the development of celluloid film, who continued to perfect the camera and projector, and who gave birth to the concept that film, outside of a theater with a stage, could tell a story in its own fashion, and who built the first sound stage studio to do it.

Thomas Alva Edison was born in Milan, Ohio in 1847. He soon was pulled out of public education and educated by his mother—although his innate technical curiosity emerged early. By the age of 10 he had established his own laboratory. As a young man, he worked for Western Electric and AT&T. One opportune day he repaired and improved a stock ticker. Grateful and cagey executives awarded him sufficient resources to establish his own shop and to hire his talented coterie. Edison died in 1931, his obituary in the *New York Times* running over 4 pages of testimony by the great and near-great. No less than Albert Einstein wrote: "A . . . great technical inventor to whom we owe the possibility of alleviation and embellishment of our outward life and all our existence with bright light."

In addition to the lights we work and live by, Edison began the projected light of the device and the aesthetic of film.

As motion pictures progressed, a farm boy in Idaho discovered how these stories of light and magic—with voice—might be sent by electronic transmission and reception, instantly as they were seen in one place to other places by electromagnetic wave radiation from cameras to Cathode Ray Tube receiver screens.[21]

Chapter 3

Tube

Philo Farnsworth

The apocryphal story goes like this: In the summer of 1921, on his parents' farm in Rigby, Idaho, he cultivated a potato field row by row, turning the soil in good true lines as any good farmer or farmer's son or daughter would. When he gazed back to survey his completed chore, the idea he had been struggling with in an instant blazed into its solution. It was clear to him at once that the electron beams that Braun's invention streamed would never produce talking moving pictures (the Holy Grail, which, he knew, motion picture technicians were pursuing). The beams had to be scanned, line by line. He was 14 or 15 years of age. When most teens struggled to comprehend what it was all about, Philo Farnsworth's genius conceived the concept to render true the most influential device of the modern and contemporary eras.

There apparently are elements of truth to this story, for, 6 years later, at the age of 21, Philo Farnsworth transmitted a televised image to a receiver. It was simply a thick white line. It would not be until the 1940's or 1950's that the device emerged ubiquitous. But it was television, and he the inventor.

Our study ends here. The trials and tribulations, the successes and patent disputes with Sarnoff and Zworykin are beyond the scope of this work.

Nonetheless, it is important to reiterate that without the discoveries of Faraday, the glass tubes of Geissler and Plucker's comprehension, Hertz's discovery of Electromagnetic Wave Radiation, Lodge's and Crooke's advances, Marconi's world-changing communications device, Braun's demonstration of castling the Cathode Ray Tube to project an image, De Forest's Audion Tube and invention of motion picture "talkies," the three men who invented film, demonstrating the visual image (and sound accompaniment) was possible, Farnsworth could not have accomplished what he did.

It wasn't merely that day on the farm, of course. He was a recognized prodigy, clarifying difficult science theories and math theorems for his classmates (and for his teachers)—Einstein's theories of relativity, and the theories behind radio and electromagnetic wave radiation—and his dream: What has become, for all its blessings and curses, perhaps the invention of most pith, moment, and impact of the technological era.

Philo Farnsworth was born near Beaver Creek, Utah in 1906. He lived long enough to see his invention beam images, sound, and voice not only vast distances on earth, but from another planet in space. He died in 1971.[22]

The year 1927 was remarkable for its haunting similarities to the events in our own time: A high water mark of wild prosperity about to crash under a President (Calvin Coolidge) who ignored a terrible Mississippi River flood in the New Orleans area, thus setting in motion his loss to Herbert Hoover in 1928. It was the year Trotsky fled the USSR as Stalin came to power. It was the year Charles Lindbergh flew the first solo non-stop trans-Atlantic flight from New York to Paris. It was the year the first trans-Atlantic telephone call was conducted from New York to London. It was the year Ford rolled off its assembly line the last Model T. It was the year Babe Ruth hit 60 home runs in one season, a record that stood for 34 years and is still most significant for the lesser number of games played. It was the year work began upon the rock of Mount Rushmore. It was the year Pan American Airways began their flights with a touch of class on board. It was the year the Holland Tunnel under the Hudson River opened, connecting by motor vehicle New York with New Jersey. And it was the year that the public recognized talking motion pictures had come to the big screen. It was the year talking moving pictures came to the small screen, albeit the public knew little or nothing about it—but it was the year of the invention that would soon bring talking moving pictures into taverns and saloons, and into our homes: The year Philo Farnsworth invented television and changed the world.[23]

Notes

[1] Baird's device was bulky. The apparatus was limited to "receivers" within close proximity. It relied on spinning disks and multiple motors stimulating one or a series of light sensitive cells. It seems to have had some success early on. The receiving sets were to be fitted with a loudspeaker. See "Televisor ..." in the *New York Times*. Not surprisingly, Pravda lays out a claim that the word "television" was coined by a Russian, an engineer named Konstantin Persky; and that the device was actually invented in 1911 by his countryman Boris Rosing. That Rosing may have produced an image using a cathode ray tube may be true; however, the unnamed researcher at bairdtelevision.com, probably protecting his namesake champion, states, "Rosing's system employed a mirror-drum apparatus as camera and a cathode-ray tube as receiver ..." Although never giving way fully to Baird's invention being more sophisticated, he does continue, giving some due credit to the Russian: "Rosing's system was primitive, but it was one of the first experimental demonstrations where the cathode ray tube was employed for the purposes of television." It is of considerable interest that Zworykin was Rosing's student and laboratory assistant. See "History of Modern Television .."; "Boris Lvovich Rosing"; and "Television."

[2] Queen Elizabeth's personal physician, William Gilbert, a curious Renaissance Man, investigated the story of Aristophanes's (the seventh century BCE philosopher, not the fifth century BCE comedic playwright) interest in generating sparks and attractions from rubbing amber with fur; Gilbert probably read Aristotle's great zoological study and so was familiar with the ancient's awareness of the electric eel. Gilbert may have begun the modern exploration, by conducting experiments showing these properties existed in many materials, not merely amber. He may have begun to uncover the relationship of magnetism to electricity.

Still, Franklin not only helped to found a nation, most scholars concur his experiments led to the intense investigation of the phenomena. As a side note, the entomology of the term derives from the Greek word for amber, *elektron*. See Sarkar, pp 7f; Cousins; Brodsky.

³This of course is a simplistic rendering of the more complex formula developed by Ohm, eponymously known as Ohm's Law, expressed as:

Where V is Voltage; I Current, R Resistance—

V generates electricity—and current, or amperage is dependent upon the R in I. Thus, in this case, V=I X R; or, to calculate R, R=V/I.

By this property of resistance, a designed element generates heat and light, as a light bulb, or amplified sound, as a radio. Thus, in the case of the light bulb, the filament within the glass enclosure contains a property which offers the current resistance. It follows, then, that the greater the Resistance, the brighter the filament will burn. Even in the ensuing and growing digital age, Ohm's Law remains constant.

⁴Sarker. See also Brodsky.

⁵Faraday published extensively. A significant amount of this primary material remains extant. An interesting opening statement in his lecture series relates in simple terms his discovery years earlier of the relationship of magnetism and electricity: ". . . There were particles of different kinds attracted to each other; and this was a great step beyond the first simple attraction . . ." See also Holmyard, *Encyclopedia* Britannica, (which Faraday himself used for research), Hamilton, and Brodsky. Dal (p. 20f) describes the very moment modern technological progress began. Lenard (p. 248f) gives a well researched, delineated observation of the experiments and their subsequent results. Here, in Faraday's own words, the moment of discovery of the light effect, now known as the Faraday effect (Einstein would later publish his paper on photovoltaic effect at the same time as his paper on the Special Theory of Relativity, another series of discoveries which changed the world forever): "Light and Electricity are two great and searching investigations of the molecular structure of bodies and it was whilst considering the probable nature and conduction and insulation in bodies not decomposable by the electricity to which they were subject, and the relation of electricity to space contemplated as void of ... matter, that (these) considerations . . . were presented to my mind" (*Experimental Researches .*, pp. 284-285).

[6]Little of Geissler's or Plucker's original notes and diaries have survived. Although their achievements have promoted enormous significance, they have been somewhat absorbed into history by the later scientists, inventors, and tinkerers, as Thomson, Lodge, Crookes, Marconi, and Edison. There are good on-line sources from reputable subscription data bases. The most significant craft and art of glass blowing has been traced to the first century B.C.(E.): "Hundreds of thin glass tubes and broken glass rods were found in a pool in the Jewish Quarter of Jerusalem ... discarded remnants of an ancient glassmaking workshop. A few small glass-blown bottles were also found . . . making this . . . the earliest example of glassblowing in the ancient world" ("What is it?"). For a succinct historical outline of glass blowing, see Thrall. For a more detailed examination of the unique craft's history, see Fischer; also see Fossing. Hittorf's revelation was heavily influenced by the work of his teacher—still it was he who recognized the rays appeared to generate from the Cathode to the Anode. The fundamental principle of the tube is as follows: When Voltage is applied to the terminals hooked to either end, an electric current flows through the tube filled with neon or argon. The current dissociates electrons from the gas molecules, creating ions; when electrons recombine with these ions, lighting effects occur. In addition to a signpost on the road to television, it should be clear that Geissler supplied the precursor for neon lights, whether gaudy, tasteful, or aesthetic.

[7]Maxwell published extensively. He read his work in public lectures. Toward the end of his life he compiled most of his derivations, calculations, deductions, and formulae in one tome, A *Treatise on Electricity* and *Magnetism*, one of the most important and significant scientific treatises ever presented. The simple title belies the profound complexity expressed within its covers. For one versed in higher mathematics it beckons a treasure trove and an insight into genius. In addition there are several good database sources. Brodsky devotes a chapter. Buchwald (*The Creation . .*) recalls Maxwell's equations and how they both influenced and challenged Hertz.

[8]Often overlooked, perchance even in this study given less attention, is another student in the brilliant panoply of pupils under Helmholtz, Eugen Goldstein. It was Goldstein who is credited with

coining the term, "Cathode Rays" for Hittorf's discovery, since, he discerned, the rays proceed from the Cathode to the Anode. Later Goldstein determined there were also Anode produced rays, which he called Canal Rays. Goldstein's work led Hertz to involve himself with Cathode Rays within Geissler's tubes. Goldstein was born in Poland in 1850, but came to Germany to study with Helmholtz. He lived until 1930, which means that before he died, he may have seen his cathode rays expressed in Farnsworth's invention. See EB, NNDB, Buchwald (*The Creation . . .*)

[9]Hertz's publication of his discoveries was not at first well received. It is a common circumstance in scientific fields for disagreements and disputes over findings to occur. But Hertz died suddenly and unexpectedly still as a young man, denying him the wherewithal to defend and to respond to his critics. Fortunately for the discoveries however, across the channel in England, within a few short years, Hertz's work was recognized for the shift in thinking and science that it was, largely by four men, three British and one Italian. Soon wireless transmission would open the twentieth century. Hertz published extensively. Most of these works remain extant. Brodsky and Buchwald provide excellent insight, and in the latter's case, the author traces each of Hertz's experiments. See also Bryant for a lucid explanation of the key experiments. *The Historical Dictionary* entry provides an adequate introduction.

[10]Considerable defense of Lodge as the inventor or precursor inventor of radio is found in his biographies. Since he lived well into the twentieth century these biographers knew him or his sons, and must therefore be taken with the overwhelming evidence supporting Marconi's invention. However, they do provide excellent insight into a full productive scientist's life, a man with one foot in the realm of pure scientific investigations and another in the ether of the arcane. See especially Jolly, and Rowlands and Wilson (*Oliver Lodge* and . . .). Larson gives an excellent review of the experiment that was at once a good introductory physics classroom show and a scientific advance that stimulated the pathway to radio. Lodge's own work, of course, should not be ignored by the serious student or scholar. It is interesting that Lodge at first considered the waves to travel through

the "ether," an unknown, invisible elemental, a flawed reasoning that also may have led to his allowing Marconi to surpass him.

For the discussion of early wireless attempts preceding Hertz and Lodge, see the "Induction Wireless" entry in *Historical Dictionary...*

[11]Unlike his colleague and friend, Crookes has not had the interest in the story of his life for a definitive biography. Gay provides one of the more detailed accounts in psychoanalysis of his challenges and techniques, and his difficult business associates. *Encyclopedia Britannica* is particularly useful. Fournier D'Albe's biography is contemporary; with a forward by Lodge, it appears quite agreeable to the scientist's life and work. *Access Science* provides a lucid and detailed explanation of the elegant breakdown of the tube designed by Crookes. Many of these discoveries often overlap. Divergent claims—founded or unfounded—are made. In this case, without question, Sir William Crookes demonstrated alone that cathode rays travel in straight lines and would phosphoresce when they strike certain materials. For nearly thirty years scientists and inventors attempted to find a way to transmit that especial light into a recognizable image.

In only a few years following Lodge's and Crookes's landmark achievements, one of the mysteries was solved. In 1897, through a series of elegant experiments using the Crookes tube and mathematical derivations, Joseph John Thomson proposed (in essence, discovered) that the stream was not molecular but sub-atomic. Indeed, Thomson identified practically the structure of the atom and certainly the whirling orbiting particle, the electron. Now it was known that Crookes's stream was an electron stream. Thomson won the Nobel Prize for Physics in 1906. He was born in Manchester in 1856. It is not hyperbolic to suggest his profound studies and results gave birth to the atomic age. Thomson died in 1940, probably aware of the evolution of the Crookes Tube into Farnsworth's and Zworykin's inventions, and most likely aware of the search for the atomic bomb. For the definitive resource concerning that search which has benefited us all and placed all of us in mortal jeopardy, see Rhodes.

[12]For an enlightening example of the snail's pace of communication and difficulty of cross-communication prior to Morse's invention, as particularly manifest by the letters between John and Abigail Adams,

especially when John, as ambassador in France and England, requested Abigail to join him, see McCullough.

The nature of scientific curiosity merely at the expense of pragmatic invention and its potential disaster is revealed as analogy to this matter before us in the intellectual exercise that ultimately resulted in the atomic bomb. Again, see Rhodes, the definitive resource.

[13]Practically unlimited resources abound for the serious student of Marconi, in no small way, the inventor, along with Edison and Tesla perhaps, who gave us the modern world. Weightman presents one of the more definitive biographies. Brodsky devotes an excellent chapter. Larson's research reveals the technical steps quite adroitly as Marconi "fined-tuned" his instrument throughout the years. The New York Times articles provide excellent reporting.

Samuel Morse was born in Charlestown, Massachusetts in 1791. A detailed recounting of his struggles to get his invention recognized and his own technical analyses of wires above the earth can be found also in Brodsky. Morse died in New York City in 1872, over twenty years before Marconi's first home tinkering attempts.

Marconi's invention had two opportunist occurrences early on, which clearly indicate in the one case that the device was so new, many were just becoming aware of it. Weightman recalls the invention being used to capture the notorious murderer Hawley Harvey Crippen, allowing Inspector Walter Dew of Scotland Yard to capture the perpetrator at sea, in a state of shock. Larson's work is particularly devoted to this chase, building to this very climactic event. Then, albeit the sinking of the *Titanic* is rife with unethical and immoral conduct, still Marconi's invention saved many lives as the *Carpathia* received the desperate message and made flank speed to rescue those who would have otherwise joined their hapless and deserted shipmates at the bottom of the sea. Later, as he maneuvered shrewdly to begin the NBC network, Sarnoff claimed he was one of the main Marconi equipment operators that terrible night, and therefore responsible for saving many lives, a claim that scholars doubt, and for which there seems little or no evidence.

[14]Several contemporary newspaper articles concerning Braun and his world-changing inventions are quite detailed and provide excellent insight that the scientist was recognized more in his

own day than years later. *Encyclopedia Britannica* and *Scientific Resource Center* provide valuable detailed information. Brodsky curiously concentrates on Braun's lesser though significant invention of the crystal diode rectifier. Larson tracks through several pages Braun's competition with Marconi and his, Braun's, considerable involvement with Telefunken, the Marconi Company's main competitor.

In any event, in the matter of the device that would herald television, that device was Braun's. Through the process of alternating voltage, it is he who first succeeded in producing the controlled stream of electrons to trace beamed patterns onto a fluorescent screen. Some of the lesser world notice that Braun suffers from today may be attributable to the fact that much about this premier scientist remains in German. Kurylo has been translated and is an excellent contemporary resource.

[15]Considerable well documented material about De Forest and his momentous inventions abound. All sources concur that De Forest was the inventor of the tube that gave radio its promotion, and of the process that gave motion pictures lip-synch sound. Levine especially provides a well documented and compelling biography of a genius in revealing the unknown to be known but fully naive in his business and patent affairs. Sadly, vulture-businessmen realized the major profits that should have been his. De Forest's ultimate satisfaction however, could never be denied him: ".. . the Audion ... a monumental invention who's (sic) magnitude is touched only two or three times in a century. The Nobel Prize winning physicist, I. I. Rabi has described it as 'ranking with the greatest of all time.' In spite of new scientific discoveries and engineering improvements the De Forest three-element audion continues to remain the foundation upon which all modern ... electronics rest."

Historical Dictionary provides an excellent sketch and lucid conceptualization of the device. In essence, De Forest added into Fleming's "valve," which had two elements, a zigzagging nickel wire, inserted between the filament and the plate. It was this element which, once the tube series received its design, amplified the signals interpreted as voice, music, and effects.

The invention then progressed rapidly. By 1920 "The results of the Warren G. Harding and James M. Cox contest, the first presidential election to be reported by radio, was broadcast from station KDKA in Pittsburgh," with a signal powerful enough in those early days that receivers up and down the East Coast could listen to the first time the results of the premier election would be known so soon. "Network radio . . . would reign supreme in reporting elections .. . until the rise of television ..." (Rasmussen).

In the early 1920's De Forest was responsible for the first lip synchronization in motion pictures by realizing that a sound strip responding to an optical reader needed placement upon the film strand itself, stimulating a calibrated photoelectric cell and together, therefore, proceeding through the projector gate. Lee De Forest, therefore, is responsible in their fashion, for two of the three most profound inventions and changes to mass communication well into the twenty-first century.

By 1927, Jack Warner had De Forest's invention usurped by wiring over 200 theatres for his production of The Jazz Singer, most commonly and erroneously studied as the first lip-synchronization film. De Forest four years earlier had demonstrated his invention in one movie house, thrilling the audience. For the lives, internecine battles, and the most significant contributions to the motion picture industry and the advance of film art, despite Jack Warner's questionable ethics, see Warner Brothers. *The Jazz Singer*, with the great Al Jolson, premiered in 1927—the same year Farnsworth invented television. The two pathways, seeking vision and speech, had converged. The one would from thence forward compliment and compete with the other.

The unnamed De Forest film that truly began lip-synch sound showed a series of vaudeville acts projected to a select audience in New York, April 1923. As *The Jazz Singer* four years later would star a vaudeville headliner at the top of his game, so did the true first film, and it was he, not Jolson, who holds the mantel of first talking in concert with lip movement in a motion picture—Eddie Cantor. It is also interesting to note that, nonetheless, *The Jazz Singer* remains the first talking motion picture to do so in three languages— English, Yiddish, and Aramaic (contrary to common understanding, the haunting Yom Kippur prelude to prayer, *Kol Nidre* (*Cancel our Vows*) is chanted on the evening of the Holy Day in the sister Semitic language to Hebrew; it remains a curious matter why Jolson, who knew all three of the Jewish languages well, eliminated the one Hebrew song from the film, Ma Tovu (*How Good _J.* Al Jolson was born in Russia in 1886 and emigrated to the U.S. with his parents at the age of seven. His all important movie truly does reflect his life, contrary to those who doubted that it was a biographical rendition. His father was a rabbi, who wished for his son (b. Asa Yoelson) to use his obvious singing talents to become a cantor; but, with Ms Americanized name, Jolson ran away from home at fifteen to join the theatre and the vaudeville circuit. And it was Ms rendition of "Mammy," although today blackface is righteously abhorrent and rightly makes us indignant, which brought him into the headliner position. He was, and the movie shows this clearly, one of the great entertainers of the twentieth century. Al Jolson died in 1950.

Eddie Cantor was born Edward Israel Iskowitz in 1892 on the lower East Side in New York City, the crowded, intense brewing pot of Jewish culture and religion in America. An orphan at the age of two, raised by his grandmother, even as a child, he entertained on the streets of the ghetto for pitched pennies. He later tried Amateur Nights at the various Yiddish theatres.

Often he won first prize. He joined the Vaudeville circuit. He captured the attention of Florenz Ziegfeld. Upon the impresario's stage he sang, danced, and downed for several years. As with most of the great Vaudeville entertainers, Jolson included, he successfully traversed into radio, and, to a lesser degree, television in its years of introduction into homes. Eddie Cantor died in 1964.

Reginald A. Fessenden was Canadian. He was born in 1866. He died in 1932. He also would have witnessed the growth of radio and the invention of television, as did Fleming.

John Ambrose Fleming was born in 1849 in England. He was knighted in 1929. He died in 1945, witnessing the astonishing changes of this study and those outside this study through nearly becoming a centenarian.

One other brilliant scientist claimed to have invented radio— Nikola Tesla. Most scholars concur that is not so. Still radio and much of modern life could not have evolved without several discoveries and developments of his—especially that of Alternating Current, an accomplishment from which all other electric and electronic inventions and advancements might not have occurred or would have occurred at a much slower pace. Even today, his design of the dynamos is what propels them to produce our current. Edison may have brought about light; but Tesla brought light to all.

Tesla was a most curious genius. He claimed to have designed a Death Ray that would destroy all enemy weapons. His attempt to generate enough power for his initial experiment burnt out the generators nearly state-wide in Colorado. Still, some scholars attest that the United States military continues to classify his design Top Secret. Nikola Tesla was born in Similjan, near or in Yugoslavia in 1856. Even as a youth, he became fascinated with the advance of electric generation. Even then he envisioned alternating current dynamos. Already in 1883 he had successfully demonstrated his concept. In 1884 he immigrated to the United States. With George Westinghouse his financier, Tesla developed the dynamos that give us our modern world, including, of course, television. Much excellent primary and secondary material on Tesla is readily available. For an introduction and thorough analysis, see O'Neill and Seifer.

Nonetheless it is important to note that Edison developed the first dynamo capable of lighting several city blocks. However, Edison could never admit that AC was superior to his beloved DC. See especially Semil.

[16]Ruspoli.

[17]Traces true photography to the invention of negative images on a paper coated with light-sensitive chemicals, by William Henry Fos Talbot in 1839. Also gives an excellent description of the *camera obscura*, the room-size device artists used to copy images in nature precisely. In essence, with its pinholes allowing the images to be formed upon one wall, the artist can trace or draw the minor up to nature exactly. With the exception of a type plate or paper plane, the room is, in effect, a large box camera.

[18]Eadweard James Muybridge was a British photographer engaged by the United States to photograph remote areas in Alaska as part of a survey team of the recently purchased territory. In 1877 he was called to California to settle a bet: Did a galloping horse ever have all four hooves off the ground?

Muybridge ... stationed twenty-four cameras side by side along a track. Twenty-four strings were stretched across the track and as the horse galloped it broke the strings and tripped the ... shutters. The result was a series of twenty-four phase pictures which ... formed a series picture ... a galloping horse does (indeed) take all four legs off the ground at one time. (Bohn et al, P.[7])

The bet was substantial. This photographic evidence made Muybridge a wealthy man. It provided the concept for the three inventors of film their impetus that photographs could, in quick series, provide the illusion of movement. Eadweard James Muybridge was born in Kingston-on-Thames in 1830. He died in 1904. See also Berdeche and Brasilliach, and the biography entry.

[19]In the 1950's a motion picture about Friese-Greene's work received some interest (***The Magic Box***). A definitive biography of his life and work continued to resurrect his heritage (Allister). See also MacCormac. Other references lay in select encyclopedia. ***The Access Science*** article is particularly useful. Friese-Greene held over sixty patents, including those in animation and color photography.

It is astonishing to think that the basic concepts of motion pictures were set down as early as 1824 by Peter Roget, the Thesaurus scholar. In 1887 Louis Aime Augustin Le Prince may very well have been the true original inventor of the camera-projector. Alas, LePrince and his invention and notes mysteriously disappeared in the same year (Whiting, p. 10). For the most exhaustive and definitive resource on the development of motion pictures before the three true inventors, see Cerum, which delineates all or nearly all of the magic lanterns, spinning disks, glass calibrated light shows, and other "Wizard of Oz" devices that thrilled parlor audiences prior to the true cameras and projectors.

[20]In 1995 40 movie directors collaborated on a film entitled ***Lumiere*** and ***Company***. All 40 shorts were shot with the Lumiere's 1895 camera; still operating as efficiently as when it was invented. Whiting presents a succinct but valuable resource. See also Bardeche and Brasillach, and Bohn et al.

[21]As might be expected voluminous material about Thomas Alva Edison, his biography, his process in the over 1,000 inventions and patents exist. Any of the encyclopedia entries give a good background. Evans provides an excellent introduction providing some detail. The author describes Edison filming Buffalo Bill and Annie Oakley, and denotes an example of the inventor's business acumen.

The general public got its first sight of the first 30-40 second films in amusement Arcades in April 1894, peering through eyepieces in an upright coin-in-the-box Kinectoscopes . . . an immediate hit. Edison sold nearly a thousand of them at $500 each ... (p. 168)

Bardeche and Brasillack detail the close dates of the several developments and achievements in film by Edison and the Lumieres. See also Bohn, et al. Edison was also responsible for advancing the process for the vacuum tube, the device De Forest improved for his invention which gave the world true radio (*Historical Dictionary*...).

David Llewelyn Wark Griffith was born in 1875 in La Grange, Kentucky, at that time a rural town about twenty miles from Louisville. His father had been a Confederate war hero, a fact which may have influenced Griffith's (wittingly or unwittingly) blatantly racist yet brilliant cinematographic landmark film, *Birth of a Nation* (1915). Griffith may be the single most important individual in the history of film. Griffith first attempted playwriting. Later, he transferred this skill to the creation of his screenplays. Only moderately successful as playwright and actor, he found his calling and the world found the art of the motion picture when he joined Edison's studio. If Porter's *The Great Train Robbery* established film as a separate art form, Griffith eventually found his way to a strand of holly trees outside of Los Angeles. With him came his coterie of actors once renowned on stage, now experienced in the fledgling and developing art of film; they were learning the various important ways in which film and film acting differs from stage preparation. He had cultivated their motion picture experience from 1903 to 1915. By 1918, the two films that established most of cinema's forms and techniques rolled out of Griffith's cameras: *Birth of a Nation* (1915) and *Intolerance* (1917). *Birth of A Nation* performed well at the box office. It performed well in expanding the art of the motion picture. However, controversy arose. The film, based on the southern sympathizer novel, The Clansman, portrayed the white sheet hooded riders as heroic cavalry rescuing white women from the fate worse than death at African-American hands. Amid the resultant charges of blatant stereotyping and racism, the previously weak National Association for the Advancement of Colored People emerged galvanized. Protests across the nation in front of the theaters began to harm receipts and riveted the country's attention at last to the injustices of the post Civil War period. For his part, Griffith protested that his offense had been unwitting.

The title of his following film demonstrates that the director understood the criticism; indeed, two of the four stories moving each in its fashion toward a climax deal with the inherent dangers of ethnocentrism. For its part, the NAACP from that moment on became a political and national force to be reckoned with, well into the 1950's, when Brown vs. Board of Education struck the second blow (President Truman had integrated the armed forces in 1948) against segregation. Although from time to time mired in controversy this valuable organization remains a viable political influence. In any event, it presents little or no neither overstatement nor hyperbole to state that Griffith developed more fully than anyone theatrical film as story telling art and contributed to the founding of Hollywood. He continued directing and making films well into the talking era. Still, as the newer directors improved upon his techniques, developing new aesthetics and technology, his stature declined. In time, the once and great motion picture revolutionary felt the art he had begun had passed him by. He died in 1948. He lies buried near his childhood home.

Interest in Griffith and his profound contribution to twentieth and twenty-first century life revived in the 1960's and 1970's as Film History studies entered the catalogues of colleges and universities. That interest continues today. Many of his films have been saved or restored. Even today, with contemporary audiences' considerable level of sophistication, the serious film student or maven who wades through the three hours of Intolerance will be rewarded with one of the most magnificent endings in all of motion pictures. For a good introduction see Drew. Gish gives a firsthand account, if biased; she was one of his favorite actresses and he her favorite director. She does well by her old good friend; it is uncertain if there was a romantic liaison; interestingly, though good friend and colleague always, in the manner of many well known actors who referred and still refer to their directors, she always called him by the honorific, "Mr. Griffith." Brown is also excellent and more objective.

For Harry Truman's landmark decision, see Sullivan; see also Staples. A good introduction for Brown (wherein is given the full title of the case, Oliver L. Brown, et. al. v. the Board of Education of Topeka (KS) et. al (1954)., and is explained with clarity how it was a combination of five cases is "Brown v. Board . . . " The NAACP web site is: www.naacp.org.; note in particular the Historical Timeline; refer to NAACP.

Lillian Gish holds the unique distinction of the longest career in motion pictures. She began with Griffith still in the New York area in 1907 or 1908. Her last picture, as she neared the age of 90, was in 1987. She died in 1996, one year short of becoming a centenarian (Osborne).

Edwin Stanton Porter was born in 1870 in Philadelphia. He worked as a telegraph operator. There is evidence he became interested in electric lights and lighting design for theatrical productions. By 1900, with this technical and aesthetic interest, Porter found his way into Edison's shop. By 1903 he had produced and directed the two films that began the art of the motion picture and no doubt stimulated Griffith on his journey: The Life of an American Fireman, which, decades ahead of his time, the director interspersed actual footage of firefighters with staged production—and *The Great Train Robbery*, which, for the first time, used film as film to tell a story. Porter went on to write, direct, or produce over 200 films, but his Edison co-worker had by 1915 established his own motion picture company in Hollywood; fans and audiences were already enthralled with and flocking to these films. Porter died in 1941.

[22]For the student or reader who wishes to pursue the inventor and his life, his wife's biography (albeit obviously biased) provides a wealth of material. A bit more objective work to begin an inquiry is Schatzkin. See also Ritchie. Brodsky provides perhaps the clearest explanation of Farnsworth's complex image dissector tube, a key component to his ultimate success (p. 142-148). Farnsworth's second transmitted image was prophetic: A dollar sign! For a fuller understanding of the backgrounds, disputes, and law suits brought against Farnsworth by Vladimir K. Zworykin and David Sarnoff, see these same resources, as well as the entries in *Historical Dictionary*. However, Evans may provide the most articulate delineation of the development and the ongoing rivalry of the two inventors and the network executive (p. 334-341).

Zworykin may very well have invented the cathode ray television system in 1923-1925. However, the patent was never granted, probably indicating the system was not fully developed. After details of Farnsworth's working device were published Zworykin visited Farnsworth's lab. He is reported to have realized what he, Zworykin, then needed to do, that is, the element, or process that had not occurred to him (perhaps the breakthrough concept of the electron wave needing to scan in both directions on the horizontal plane). Nonetheless Zworykin is a huge figure in the development of television. Many of the improvements for the device, more fully developed by the late 1940's, including the camera, transmitter, and receivers, largely as we recognize the system today, along with Farnsworth's ongoing contributions to the devices, were influenced or improved upon by him. Nonetheless evidence is strong that the invention ineffably was Philo Farnsworth's. "Farnsworth envisioned a system that electronically scanned each scene line by line and recreated the image line by line at the receiver... (This) required an electronic camera he called the 'image dissector.' The idea was to focus the image on a light-sensitive plate and scan the plate line by line; the resulting electronic signal could be transmitted and used to recreate the scene line by line at the receiver on Farnsworth's 'image oscillate,' (the) cathode ray tube ..." (Brodsky, p. 142-143). In other words, television!

Vladimir Kosma Zworykin was born in Murom, Russia, in 1889, into a wealthy family. As with other scientists we have reviewed, his propensity for science and electronics became manifest at an early age. By 1910 the young scientist was already interested in the chase to invent "visual wireless," i.e. transmitted and received electronic television. Zworykin served the motherland during World War I, where he was most needed, in the Signal Corps. Following the mysterious disappearance of his mentors and colleagues during the Bolshevik Revolution, Zworykin immigrated to the United States. There has been a persistent rumor that he was helped by George Westinghouse. Little evidence validates the rumor. It is known that he worked in the Russian embassy for a while, probably due to his proclivity with languages.

As indicated his investigations to develop electronic tubes for camera and receiver, that is, a working television system, in the early 1920's paralleled Farnsworth's. The two did not at first realize they were in this "race." Zworykin kept up with publications in the field. He recognized that Farnsworth had conquered the essence of the problem. Still, his own kinescope (receiver) and iconoscope (camera), as he called them, with the support of RCA (he was appointed the Director of the Electronic Research Laboratory), allowed him greater resources to advance the invention at a faster pace than Farnsworth's company. By 1933, he had developed the complete system with a resolution fire of 240 scanned lines (today's televisions scan nearly 2,000 lines a second). Farnsworth kept up, but eventually settled the patent dispute as described. By the late 1930's and into the 1940's, Farnsworth's and Zworykin's improvements resulted in television largely as we know it, ready for mass consumption, studio broadcasts, and all its greatness and ills. Vladimir K. Zworykin died in Princeton, New Jersey in 1982. See "Vladimir Kosma Zworykin," an excellent chem..ch.huji.ac.il article. Also see Zworykin's own publications. Amore thorough biography is Abramson. See also Cheek and Kim.

David Sarnoff was born in 1891 in Russia. As a child, he studied Torah and Talmud, his parents designating him for the rabbinate. Upon entering the United States in 1900, however, the boy determined he would learn English and succeed in the secular world of the American dream. Early in this pursuit he studied with Marconi; he operated one of the company's transmitters and receivers. By the 1920's he saw the vast potential of radio. Founding RCA and NBC, he remained the president and motivating force through the development of television well into the 1970's. Begrudgingly recognizing Farnsworth had invented television and not his own engineer, a fellow Russian he claimed he and not Westinghouse had brought over, Sarnoff kept up a barrage of patent fights, ultimately settling with Farnsworth for one million dollars for perhaps the most influential invention of the twentieth and twenty-first centuries. David Sarnoff died in New York at the end of 1971. See Bilby; also the New York Times obituary, and Historical Dictionary ...

[23]"1927 History."

Selected Bibliography

Abramson, Albert. Zworykin, Pioneer of TV. Foreword Erik Barnouw. Champaign: University of Illinois Press, 1995.

Allister, Ray. Friese-Greene: Close-up of an Inventor. New York: Arno Press, 1972.

Andrews, Peter, in Oliver Lodge and the Invention of Radio, which see. Bardeche, Maurice and Robert Brasillach. The History of the Motion Pictures. New York: W. W. Norton and Company, Inc. and The Museum of Modern Art, 1938.

Bellis, Mary. "Philo Farnsworth." inventors.about.com, "The History of the Cathode Ray Tube. Ibid. 2007.

Bilby, Kenneth. The General. David Sarnoff and the Rise of the Communication Industry. New York: Harper and Row, 1986.

Bohn, Thomas W. Richard L. Stromgren and Daniel H. Johnson. Light and Shadows: A History of Motion Pictures. 2nd ed. Sherman Oaks: Aefred Publishing Company, Inc., 1975.

"Borisvovich Rosing." http://www.bairdtelevision.com/rosing. html. "Bose, Sir Jagadis Chandra." Encyclopedia Britannica. EB Online. Ret. June 5, 2008.

Bragg, Sir William H. "Met Who Put Science to the Uses of Mankind." New York Times. June 22, 1930, xx, 3.

"Braun, Karl Ferdinand (1850-1918)." Access Science. www. access science.com.library.

"Braun, Ferdinand." Encyclopedia Britannica. EB Online. Ret. June 5, 2008. Braun, Karl Ferdinand. "Electrical Oscillations and Wireless Telegraphy." Nobel Lecture, December 11, 1909. http:// Nobelprize .org/Nobel prizes/ physics/laureates.html.

Braun, Ferdinand. Uber Drahtlose Telegraphie and Neure Physikalische Forschungen. Strassburg: J. Heitz and Mundel, 1905.

Brodsky, Ira. The History of Wireless. St. Louis: Telescope Books, 2008. Brown, Karl. Adventures with D. W. Griffith. New York: Farrar, Straus and Giroux, 1973.

"Brown v. Board of Education: About The Case." Brown Foundation for Educational Equity, Excellence and Research. http:// brownvboard.org., rev. April 11, 2004.

Bryant, John H. "Heinrich Hertz's Experiments and Experimental Apparatus: His Discovery of Radio Waves and His Delineation of

Their Properties." Heinrich Hertz: Classical Physicist, Modern Philosopher. Ed. Robert S. Cohen and Max W. Wartofsky. Dordrecht: Klurver Academic Publishers, 1998, pp. 39-58.

Buchwald, Jed. The Creation of Scientific Effects: Heinrich Hertz and Electric Waves. Chicago: The University of Chicago Press, 1994.

_____."Reflections of Hertz and the Hertzian Dipole." In Bryant, which see, pp. 269-280.

"Cantor, Eddie." Wilson Web. http://o-vnweb.hwwilsonweb.com. From Current Biography, 1954.

"Cathode Ray." Science Resource Center. http://0-galenet.galegro up.com. Ret. September 18, 2008.

"Cathode-Ray Tube." Science Resource Center. http://0-galenet .galegroup.com. Ret. September 18, 2008.

Ceram, C. W. Archaeology of the Cinema. New York: Harcourt, Brace & World, Inc. c. 1935.

Cousins, Margaret. Thomas Alva Edison. New York: Random House, 1993.

_____ Ben Franklin of Old Philadelphia. New York: Random House, 1980.

"Crookes and Geissler Tubes." www.sparkmuseum.com/glass. htm. "Crookes, Sir William." Encyclopedia Britannica. EB Online. Ret. June 5, 2008.

"Crookes, William (1832-1919)." Access Science. Accessscience .com. Ret. July 3, 2008.

Dahl, Per F. Flash of the Cathode Rays: A History of J. J. Thomson's electron. Bristol: Institute of physics Publishing, 1997.

"Dawn of the Electrical Age, The." Sparkmuseum.com/Highli ghts.htm. "David Sarnoff is Dead: Visionary Broadcast Pioneer." New York Times, December 13, 1971. Obit., pp. 1, 43.

Doucel, Manuel G. "On Hertz's Conceptual Conversion: From Wire Waves to Air Waves." In Bryant, which see, pp. 73-88.

Douglas, Susan. Inventing American Broadcasting, 1899-1922. Baltimore: Johns Hopkins University Press, 1987.

Drew, William M. "D.W. Griffith (1875-1948)." www.gildasattic. com. "Dr. Ferdinand Braun Dead." New York Times, April 22, 1918, p. 11.

Dunlop, Orrin E. Marconi: The Man and His Wireless. New York: Macmillan, 1937.

"Edison is mourned as Leader of Age." New York Times. October 19, 1931, p. 24.

Edison The Man. Motion Picture. Spencer Tracy, Rita Johnson, Charles Coburn. Screenplay Bradbury Foote. Dir. Clarence Brown. Rel. May 10, 1940.

"Edison, Thomas Alva (1847-1931)." Access Science. www. access science.com. Ret. July 30, 2008.

Einstein, Albert. Special Theory of Relativity. 1905. The Albert Einstein Home Page. www.Humbold. 1 .com.

"Electromagnetism." Encyclopedia Britannica. Encyclopedia Britannica online. Ret. July 30,2008.

"The Ether is Discarded." New York Times. June 15, 1930, p. xx, 9. "Eugen Goldstein." nndb.com. ret. August 14, 2008.

Evans, Harold, Gail Buckland and David Lefer. They Made America. New York: Little, Brown And Company, 2004.

Everson, George. The Story of Television; The Life of Philo T. Farnsworth. New York: W. W. Norton, 1949.

Fahie, John. A History of Wireless Telegraphy, Including Some Favorable Proposals for Subaqueous Telegraphs. Edinburgh: Blackwood, 1899. Fantel, Hans. "Sound; 80 Years Ago This Month, It Was Live From The Met." New York Times: June 24, 1984, p. A23.

Faraday, Michael. Correspondence: 1832-December 1840. Cont. Frank A. J. L. James. 1 ET, 1993.

_____. "Lecture V: Magnetism-Electricity." A Course of Six Lectures on the Various Forces of Matter and Their Relations to Each Other.

Ed. William Crookes. Delivered Before a Juvenile Auditory at the Royal Institution of Great Britain During The Christmas Holidays of 1859-1860. New York: Harper & Bros., 1868, pp. 120-144.

_____. Experimental Researches in Electricity. Vol. ii. London: Richard and John Edward Taylor, 1844, pp. 284-285.

Farnsworth, Elma. Distant Vision: Romance and Discovery on an Invisible Frontier. Salt Lake City: Pemberly Kent Publishers, 1990.

"Farnsworth, Philo Taylor." Encyclopedia Britannica. Encyclopedia Britannica Online. Ret. June 5, 2008.

Fischer, Alysia. Hot Pursuit: Integrating Anthropology in Search of Ancient Glassblowers. Lanham: Lexington Books, 2007.

Fisher, David E. and Marshall John Fisher. Tube: The Invention of Television. San Diego: Harcourt Brace & Company, 1996.

Fossing, Poul. Glass Vessels Before Glass-blowing. trans. W. E. Calvert. Copenhagen, 1940.

Fourner d'albe, Edmund Edward. The Life of Sir William Crookes, 0.M., F. R. S. Foreword mSir Oliver Lodge. New York: Appleton & Company, 1924. "Friese-Greene, William (1855-1921)." Access Science. Accesscience.com. ret. July 20, 2008.

"Friese-Greene, William." Encyclopedia Britannica. Encyclopedia Britannica Online. Ret. July 30, 2008.

"Future of Wireless Telegraphy." New York Times: May 7, 1899, p. 20. Hamilton, James. A Life of Discovery: Michael Faraday, Giant of the Scientific Revolution. New York: Random House, 2002.

Hawks, Ellison. Pioneers of Wireless. London: Methuen &Company, 1927.

Hertz, Heinrich. Electric Waves: Being Researches on the Propagation of Electric Action With Finite Velocity Through Space. Trans. D. E. Jones. Pref. Lord Kelvin. New York: Macmillan (St. Co., 1893.

_____. "On the Relations Between Maxwell's Fundamental Electromagnetic Equations and the Fundamental Equations of the Opposing Electromagnetics." Trans. Unknown. Annual Physical Chemistry, 23:84-103, 1884.

"Hertz, Heinrich Rudolf (1857-1894)." Access Science. www. acces science.com.

Hertz, J. Heinrich Hertz: Memoirs, Letters, Diaries. 2nd ed. Trans. L Brinner, M. Hertz and C. Susskind. San Francisco: San Francisco Press, 1977.

Historical Dictionary of American Radio. Ed. Donald G. Godfrey and Frederic A. Leigh. Westport: Greenwood Press, 1998.

"History of Modern Television Started in Russia in 1900." Pravda. August 18, 2005, English Tr. Online: http://english.pravda.ru/science/tech/8778-television-0.

Holmyard, E. J. "Michael Faraday (1791-1867)," p. 45f; "James Clerk Maxwell (1831-1879," p. 54f; "Sir William Crookes (1832-1919," p. 59f; "The Structure of the Atom," p. 73. British Scientists. New York: The Philosophical Library, Inc., 1951.

Hong, Sungook. Wireless: From Marconi's Black Box to the Audion. Boston: Massachusetts Institute of Technology Press, 2001.

Hunt, Bruce J. "Engineering and Applied Sciences—The Creation of Heinrich Hertz and Electric Waves." Rev. of J. Z. Buchwald. American Scientist. Vol. 83, iss. 6, p. 584, Nov. 1995. listed, G. A.

"Guglielmo Marconi and the History of Radio." Pt. 1 and 2. GEC Review 7, no. 1, 2, 1991.

"Inventor: Edison." Over and Under Rated. American Heritage. October, 2003, pp. 44-45.

"James Clerk Maxwell." Studyworld Biography: Historical Figures. http:// studyworld.com. ret. June, 2008.

"Johann Heinrich Wilhelm Geissler." World of Invention. Ed. Kim L. Detroit: Thomson Gale, 2006. Science Resource Center.gale. www.galenet.Galegroup.com. ret. July 30, 2008.

Jolly, W. R Sir Oliver Lodge. London: Constable and Co. Ltd., 1974.

"Jolson, Al." Wilson Web. http://)-vnweb.hwwilsonweb.com. From Current Biography, 1954.

"Joseph John Thomson." www.chemheritage.org. ret. September 18, 2008. "J. J. Thomson." www.nobelprize.org. ret. September 18, 2008.

"Karl Ferdinand Braun." Science Resource Center. http://0-gale net.galegroup.com. Ret. September 18, 2008 (Articles 1/18 and 2/18).

Knight, David M. Science in the Romantic Era. Brookfield: Ashgate/ Variorum, 1998.

Kurylo, Friedrich. Ferdinand Braun: A Life of the Nobel Prize Winner and Inventor of the Cathode-Ray Oscilloscope. Trans. Charles Susskind. Pub. In German, 1909.

Larson, Erik. Thunderstruck. New York: Crown Publishers, 2006.

Lenard, Phillipp Eduard Anton and H. Hatfield. Great Men of Science: A History of Scientific Progress. New York: Macmillan Company, 1933. Lessing, Lawrence. Man of High Fidelity: Edwin Howard Arm- strong. New York: J. B. Lippincott, 1956.

Lewis, Tom. Empire of the Air: The Men Who Made Radio. New York: Edward Burlingane Books, 1991.

Levine, I. E. Electronics Pioneer: Lee De Forest. New York: Julian Messner, 1964.

"Lodge, Sir Oliver Joseph." Encyclopedia Britannica. Encyclopedia Britannica Online. Ret. June 5, 2008.

Lodge, Sir Oliver. "The Luminiferous Ether and the Modern Theory of Light." The Ether of Space. New York: Harper Brothers, 1909.

_____. Past Years. Hodder & Storeghton, 1926.

_____. Pioneers of Science and the Development of Their Scientific Theories. New York: Dover Publications, 1960.

_____. Signaling Through Space Without Wires: Being a Description ofHertz and His Successors. London:

"The Electrician" Printing and Publishing Company, Ltd., 1894.

"Lumiere." Access Science. Asscesscience.com. ret. July 30, 2008.

The Magic Box. Motion Picture. Robert Donat, Margaret Johnston, Maria Schell. Screenplay by Eric Ambler. Dir. John Boulting. Rel. September 22, 1952.

Maccormac, John. "British Film History Recalled by Pioneers." New York Times: September 30, 1928, p. 121.

"Marconi, Guglielmo." Encyclopedia Britannica. Encyclopedia Britannica Online. Ret. June 5, 2008. "Marconi's System Explained." The London Truth. Prob. May, 1899. See New York Times: May 14, p. 6.

Marconi, Guglielmo. Transatlantic Wireless Telegraphy. Lecture given before the Royal Institution, Friday, March 13, 1908.

_____ "Wireless Telegraphic Communication." Nobel Lecture, December 11, 1909. http:Nobelprize.org/nobelprizes/physics/laureates/1909/ Marconi-lecture.html.

_____. Wireless Telegraphy: Lecture Given Before the Liverpool Chamber of Commerce, Monday, 24 February, 1908. Marconi Wireless Telegraph Company, 1908. "Marconi's Telegraphy." New York Times: January 23, 1898, p. IWM3.

Mashelker, R. A. "Ahead of the Curve: J. C. Bose Believed that by Focusing on the Boundaries Between Different Sciences, He Would Be Able to Demonstrate the Unity of All Things. India Today, April 21, 2008. "Maxwell, James Clerk (1831-1879)." Access Science. Accessci- ence.com. ret. July 30, 2008.

Maxwell, James Clerk. A Treatise on Electricity and Magnetism. Vol. 1. New York: Dover Publications, 1891. Repub. 1954.

McChesney, Robert W. "Lee De Forest and the Fatherhood of Radio." The Journal of American History. Bloomington: V. 81, Iss. 1, June 1994, p. 310. McCullough, David. John Adams. New York: Simon & Schuster, 2001. Mervis, Jeffrey. "Bose Credited With Key Role in Marconi's Radio Breakthrough." Science. Vol. 279, Iss. 5350, January 23, 1998, p. 476. Mufson, Steven. "Empire of the Air." Review. The Washington Post, February 9, 1992.

"Muybridge, Eadweard James." Entry, A Dictionary of Scientists. Oxford: Oxford University Press, 1999.

NAACP. www.naacp.org.

"New Cathode Ray Tube Pushes Television Ahead." New York Times: November 24, 1929, p. xx, 12.

Nichols, Edward Nimingtoy, H. L. Howes and Dt. T. Wilber. Cathodo-Luminescence And the Luminescence of Incandescent Solids. Washington:

The Carnegie Institute of Washington, 1928.

Nier, Keith A. "Zworykin, Pioneer of Television. The American Historical Review. Vol. 102, Iss.3, June 1997, p. 913. "1927 History." www.thepeoplehistory.com/1927.html. The New York Times Archives. Nikola Tesla: Lecture Before the New York Academy of Sciences, April 6, 1897. Reconst. Ed. Leland I. Anderson. Incl. ed. Remarks; Introduction; Background: "Section I: Adaptive Wireless Telegraphy Receiving Methods." Breckenridge: Twenty-First Century Books, 1994.

Oliver Lodge and the Invention of Radio. Ed. Peter Rowlands and J. Patrick Wilson. Liverpool: PD Pub's, 1994.

O'Neill, John J. Prodigal Genius: The Life of Nikola Tesla. New York: Ines Washhern, Inc., 1944.

Osborne, Robert. "Commentary, Orphans of the Storm." Dir. D. W. Griffith. With Lillian Gish, 1923. Turner Classical Movies, November 25, 2008. "Plucker, Julius (1801-1868)." Access Science. Accesscience.com. ret. June, 2008.

"Possibilities of Electricity." New York Times: February 21, 1892, p. 4. Report of W. Crookes's article from his lecture in Fortnightly Review, prob. January.

"Professor Wright on Cathode Rays." New York Times, February 16, 1896, P. 9.

"Radio Development in the 1900's, 1900-19090. Gale Research, 1997. http//:galenet.Galegroup.com.

"Radio." Encyclopedia of American Cultural and Intellectual History. Three vols. New York: Charles Scribner's Sons, 2001.

Gale Encyclopedia of United States Economic History; http://galenet.galegroup.com, 1999.

Rasmussen, Frederick N. "When Wilson Beat Hughes, Baltimore Blinked." Back Story. The Baltimore Sun. November 2, 2008: Maryland Closeup, p. 5. Rhodes, Richard. The Making of The Atomic Bomb. New York: Simon & Schuster, 1986

Ritchie, Michael. Please Stand By: A Prehistory of Television. Woodstock: The Overlook Press, 1994.

Rowlands, Peter. "Radio Begins in 1894": In Oliver Lodge and the Invention of Radio, which see.

_____. "Radiowaves":

_____. "Waves From the Sun"

Ruspoli, Mario. The Cave of Lascaux: The Final Photographs. Pref. Yves Copperns. New York: Harry N. Abrams, 1986.

Sarkar, Tapan K. et. Al. History of Wireless. Hoboken: John Wiley & Sons, Inc., 2006.

Schatzkin, Paul. The Boy Who Invented Television.?: TeamCom, 2002.

_____. The Farnsworth Chronicles. Farnovision.com/chronicles, 1977, updated 2001.

Sealey, David. "Marconi Waves": In Oliver Lodge and the Invention of Radio, which see.

Seifer, Marc J. Wizard: The Life and Times of Nikola Tesla: Biography of a Genius: Seacaucus: Birch Lane Press, 1996.

Settel, Irvin. A Pictorial history of Radio. Int. Brock Brower. New York: Crosset & Dunlap, 1960.

Shales, Tom. "Radio's Fathers of Invention: 'Empire'; Dark Birth of Broadcasting." The Washington Post: January 29, 1992, b.01.

"Simplified receiver for Television Shown." New York Times: November 19, 1929, p. 30.

"Sir Joseph John Thomson." www.phy.hr/. ret. September 18, 2008.

Smil, Vaclay. Creating the Twentieth Century: Technical Innovations of 1867-1914 and Their Lasting Impact. Oxford: Oxford University Press, 2005.

Sontag, Susan. On Photography. New York: Farrar, Straus and Giroux, 1977. Sporre, Dennis J. "Pictures, 'Photographic Techniques,' Chapter 2. Perceiving the Arts: An Introduction to the Humanities. 9th ed. Upper Saddle River: Pearson: Prentice Hall, 2009.

Staples, Brent. "Gay Bigotry." The New York Times: Editorial, November 8, 1999.

Sullivan, Ronald. "The Gay Troop Issue; The Military Balked at Truman's Order, Too." The New York Times: Editorial, January 31, 1993.

"A Talk by Nikola Tesla." New York Times. May 24, 1891.

"Television." Encyclopaedia Britannica. B. Online Academic Edition. "Televisor Lets Radio Fans 'Look in' As Well As Listen." New York Times: April 25, 1926, pp. xx, 17.

"Thomson, J(oseph) J(ohn) (1856-1940)." Access Science. Accesscience.com. ret. July 30, 2008.

Thrall, Anne. "A Brief History of Glass Blowing." www.neder. com / glassact/history.html.

"Titanic Accident and the Radio Act of 1912." Discovery United States History. Gale Research, 1997. http://galenet.galegro up.com.

"The True Cathode Rays." The New York Times: May 24, 1891.

Turney, Walter J. "Professor Braun's Experiments in Directing Wireless Messages." The New York Times: September 3, 1905, p. SM, 4.

"Urges Faraday Tribute." The New York Times: November 18, 1929, p. 27. "Vladimir Kosma Zworykin." www.chem.ch.huji.il/history. Ret. November 4, 2008.

"Vladimir Zworykin—Electronic Television System." www.inventors.about.com. Ret. November 4, 2008.

Warner Brothers. Documentary Film. Public Broadcast System. Nar. Clint Eastwood: September, 2008.

Weightman, Gavin. Signor Marconi's Magic Box. Cambridge: Da Capo Press, 2003.

"What is It?: Glass Refuse." Biblical Archaeology Review. V. 34, No. 6, Nov/ Dec 2008, pp. 24, 80.

Whiting, Jim. Auguste & Louis Lumiere and the Rise of motion Pictures. Hockessin: Mitchell Lane Publishers, 2006.

Wilson, J. Patrick. "Oliver Lodge: A Sketch of His Life." In Oliver Lodge and the Invention of Radio, which see.

"Wireless Devices Set Up in Court." The New York Times: May 7, 1915, p. 8. "Wireless Telegraphy Test." The New York Times: April 27, 1899, p. 7. "WQXR Marks 60th Year With a Series on its Past." The New York Times: December 4, 1996, p. C 14.

Ybarra, T. R. "Marconi Recounts Birth of Wireless." The New York Times: December 11, 1926, p. 1.

Zworykin, Vladimir K. and E. D. Wilson. Photocells and Their Applications. New York: Wiley, 1930.

Zworykin, Vladimir K. and G. A. Morton. Television: The Electronics of Image Transmission. New York: Wiley, 1940.

Index

About the Author

Donald Ray Schwartz has published nearly 200 works, including essays, articles, reviews and criticisms, a novella, and non-fiction works. Lillian Russell: A Bio-Bibliography, in collaboration with Anne Bowbeer is considered the definitive resource on the late 19th, early 20th centuries chanteuse and a significant contribution to that period of American theatre in general. Noah's Ark: An Annotated Encyclopedia of All The Animal Species in the Hebrew Bible was the Jewish Book Club Selection for the year it was published and is still considered the definitive resource for that subject. Professor Schwartz's plays have been produced up and down the east coast and throughout the Midwest; Review won the Sarasota (Florida) Theatre National Playwriting Contest. His poem, *The Cross Country Run of Jennifer X. Dreifus*, won the Mellen National Epic Poetry Contest. Professor Schwartz has directed or produced over 40 main stage productions (including full stage musicals). He has directed television commercials. He has featured cameo roles in 2 independent motion pictures. He was a featured performer for Nebraska Public Television's industrial film series. Having studied intensively with several prominent rabbis in their fields of scholarship, Professor Schwartz presents a series of lectures for the CCBC Speaker's Bureau, the college's conferences, and in the community with the theme, "Rhetoric and the Hebrew Bible." Donald Ray Schwartz is Associate Professor of Speech, Theatre, and Mass Communication at The Community College of Baltimore County.

He resides in Baltimore, with his wife Ann.

www.ingramcontent.com/pod-product-compliance
Lightning Source LLC
Chambersburg PA
CBHW071513210326
41597CB00018B/2740

* 9 7 8 1 9 6 6 7 8 2 3 2 2 *